Work 135

面向同一方向

Facing in the Same Direction

Gunter Pauli

[比] 冈特·鲍利 著

[哥伦] 凯瑟琳娜·巴赫 绘

高青 译

上海远东出版社

丛书编委会

主　任：田成川

副主任：闫世东　林　玉

委　员：李原原　祝真旭　曾红鹰　靳增江　史国鹏

　　　　梁雅丽　孟小红　郑循如　陈　卫　任泽林

　　　　薛　梅　朱智翔　柳志清　冯　缨　齐晓江

　　　　朱习文　毕春萍　彭　勇

特别感谢以下热心人士对童书工作的支持：

匡志强　宋小华　解　东　厉　云　李　婧　庞英元

李　阳　梁婧婧　刘　丹　冯家宝　熊彩虹　罗淑怡

旷　婉　王靖雯　廖清州　王怡然　王　征　邵　杰

陈强林　陈　果　罗　佳　闫　艳　谢　露　张修博

陈梦竹　刘　灿　李　丹　郭　雯　戴　虹

目录

Contents

ZERI Learning Initiative

数百只鹈鹕正聚在一起去觅食，一大群羊正顺着河游泳。一条凯门鳄和另一条短吻鳄一直跟着它们，随时寻找时机袭击它们。

"你瞧它们，"凯门鳄说，"所有这些鹈鹕全朝着同一个方向行进，偷袭它们简直是易如反掌啊。"

Hundreds of pelicans are gathering to go feeding. The large flock of sheep is swimming down river. A caiman and an alligator are following them, looking for the best way to ambush them.

"Look at this," says the caiman. "All those pelicans are facing in the same direction. It should be very easy to sneak up on them."

数百只鹅鹕正聚在一起……

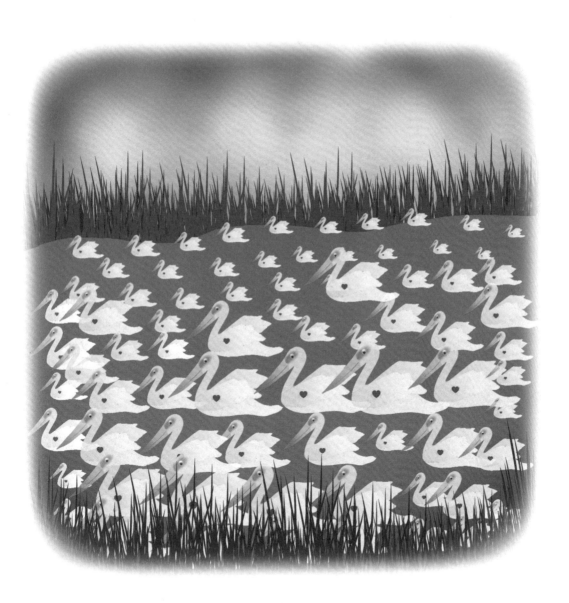

Hundreds of pelicans are gathering ...

它们根本就没有注意到我们的存在。

They don't even notice that we are here.

"它们根本就没有注意到我们的存在。"短吻鳄说，"它们简直和人类的行为一模一样，盲目地随大流，直到掉进我们的陷阱时，都还不明白发生了什么！"

"也就有那么两三只可能会朝另一个方向行进。"

"是的，但没有人会注意到这几位的行为，没准其他人还会认为它们是另类呢。"

"They don't even notice that we are here," comments Alligator. "They are behaving just like humans. They do not understand what is happening. They just blindly follow the crowd until right into our trap!"

"There are perhaps two or three looking in the other direction."

"That is true, but no one pays any attention to them. The others probably think those ones are just eccentric."

"快瞧，那边把它的头扎在水里的那位。"

"典型例子！其实它意识到了潜在的危险，但它不想去了解这危险是什么，因为它认为受害者会是别人，而不是它自己。"

"真是令人惊异，即使这些鹈鹕知道它们正在朝着错误的方向行进，而且是径直游向我们的陷阱，也没有一只鹈鹕改变行进的路线。"

"And look at that one over there, sticking its head in the water."

"Typical! It knows there is danger lurking, but does not want to know anything about it. It thinks everyone else might be a victim, just not him."

"How amazing that even when these pelicans know that they are heading in the wrong direction, swimming straight into our trap, no one is changing course."

......把它的头扎进水里。

... sticking its head in the water.

爱因斯坦把这称之为 "愚蠢"。

Einstein called it "stupidity".

"和人类一样啊，他们知道会面临一个即将发生的危机，但他们还是继续做同样的事情，期待着会有不同的结果发生。你知道，爱因斯坦把这称之为'愚蠢'。"

"或者，他们只是事后解释一下所发生的事情，舔舔伤口。但这于事无补。"

"The same with people that they know they have a looming crisis, and yet they continue to do the same things, expecting different results. You know, Einstein called it 'stupidity'."

"Or, they only explain what had happened afterwards, while licking their wounds. But that does not change anything."

"人类和鹈鹕似乎只是都简单地认为：既然受伤害的是少数，就无须未雨绸缪。他们为什么毫不关心其他同胞的幸福呢？"

"什么意思？"

"嗯，举个例子，人们常说他们要避免危机、解决问题，却让众多的年轻人失去工作。"

"People and pelicans both seem to simply accept that some will get hurt, and will have no more purpose in life. Why are they so unconcerned about the well-being of their fellows?"

"What do you mean?"

"Well, for instance, people say they want to avoid a crisis and solve a problem, but then leave so many young people without jobs."

...... 其他同胞的幸福呢?

... well-being of their fellows?

......都是他们父母已经知道的。

... everything their parents already know.

"毫无道理呀。为什么让社会上那些活力十足的年轻人无工作可做呢？"凯门鳄问。

　　"更糟糕的是，他们让年轻人认真学习和研究的都是他们父母已经知道的。"

"This makes no sense. Why would you leave the youngest and the most energetic members of society without work?" Caiman asks.

"What is even worse, they let their young people study and learn by heart everything their parents already know."

"但如果仅仅学习父母的经验，他们怎么避免危机或解决问题呢？" 凯门鳄问。

　　"孩子们的想象力远超他们的父母。"

"But then how can they ever avoid a crisis or solve a problem if they only study what their parents know?" Caiman asks.

"Children are able to imagine things their parents can't."

The Blue Economy

......它们需要朝着另一个方向行进。

... they need to look in the other direction.

"你说得对，它们需要朝着另一个方向行进。希望这些鹈鹕行动时没有注意到我们。"

　　短吻鳄说："我们的孩子们应该做些父母们从未做过的事情。"

　　"我们知道所有的父母都希望自己的孩子能够超越他们。"

"You are right, they need to look in the other direction. Let's hope these pelicans do not notice us when they do."

"Our children should be looking for something their parents never looked for." says Alligator.

"We know all parents want their children to do better than they have done."

"但是，如果年轻人只想得到问题的现成答案，怎么能让人去学习一些新的东西呢？"

"没错！哦，我真期盼着有一天，我们能问出还没有已知答案的问题，这样我们就可以解决社会面临的实际问题。"

"这让我对未来燃起了希望！"

……这仅仅是开始！……

"But then, how can anyone learn something new when young people are only expected to answer questions for which everyone already has the answer."

"Exactly! Oh, I am looking forward to the day when we can ask questions, to which no one has an answer. In this way we will solve the real problems of our community."

"Now that gives me hope for the future!"

... AND IT HAS ONLY JUST BEGUN! ...

… AND IT HAS ONLY JUST BEGUN! …

The mathematics and the logic of flocking birds and swarming insects have been used by airlines to simulate passengers boarding a plane and to assign arrival gates for aircraft at airports.

航空公司已经利用鸟群和昆虫群的数学逻辑原理，来模拟乘客登机，以及为机场的飞机分配到达口。

Testing shows that human crowds can predict individuals across a variety of real-world projections.

测试表明，人群可以通过现实世界中的各种行为来对个体的行为进行预测。

A caiman has a rounded snout and dagger-like teeth. An alligator has conical shaped teeth and a beige mouth colour.

凯门鳄有一个圆圆的鼻头和匕首般的牙齿。短吻鳄具有圆锥形的牙齿和米色的嘴唇。

The pelicans fish by swimming in line or forming a "u" shape, and by beating their wings on the water surface they drive fish into shallow water where they can easily scoop them up.

鹈鹕们捕鱼时会排成一条直线或是 U 型，用翅膀扑打水面，将鱼驱赶至浅水区，在那里它们可以轻松地用嘴把鱼"舀"起来。

The fisherman let the pelicans do not store fish in their pouch, but simply use it to catch other fish and then grab their heads before they immediately swallowing the fish.

渔民不让鹈鹕把抓到的鱼吞进喉囊里，而只是用它来捕获其他鱼，然后就把鹈鹕的头掐住，以便在其吞咽之前将鱼吐出。

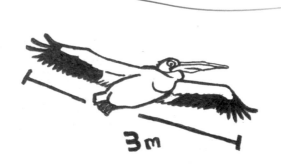

3m

The pelicans have a wingspan of up to 3 metres, and can fly at altitudes of up to 3,000 metres. To aid them in swimming and improve aerodynamics in flight, they have air pockets located between their bones, under their skin and wings and also in their chest and neck.

鹈鹕的翼展可达3米，可以在高达3 000米的高度飞行。它们的骨骼之间、皮肤和翅膀之下以及胸部和颈部都有气囊，可以帮助它们游泳，提高在空气中飞行的动力。

Herd instinct is characterised by a lack of individual decision-making, and people think and act in the same way. In the case of investments, people buy or sell shares and stocks based on the fact that many others are investing or disinvesting in those stocks.

缺乏个人决策是群体本能的特点，人们也是以同样的方式思考和行动。例如投资时，人们会根据许多其他人对股票的投资情况来购买或出售股票。

It is a myth that ostriches bury their head in the sand when they are scared or threatened. Ostriches scrape holes in ground as nests for their eggs. They will stick their head into the nest to turn the eggs and this action may have given rise to the myth.

有一种虚构的说法，鸵鸟被吓到或受到威胁的时候会将头埋进沙子里。鸵鸟在地面挖洞作巢来孵蛋。它们把头部伸进窝里来转蛋，这个行为可能导致了上述说法的出现。

If we only learn what our parents know, will we be able to do better than they have done?

如果我们仅仅学习我们的父母所知道的知识，我们怎么能做得比他们更好呢？

If there is a problem, would you hide away and hope that others could solve it?

如果碰到问题，你会隐藏起来，希望别人来解决吗？

Do you enjoy exams that the teacher already knows the answers to all the questions. Or do you prefer questions to which no one has the answer?

你是喜欢参与老师已经知道所有问题答案的考试，还是更喜欢研究没有人知道答案的问题？

When everyone is blindly following the crowd, are you one of those just being swept along by them? Or are you alert and aware, checking and verifying whether the group is moving in the wrong direction?

当每个人都盲目随大流时，你会是其中的一员吗？还是说你会保持警惕和清醒，去检查和验证该群体是否走错了方向？

Do It Yourself!
自己动手!

When you go to a football game or any sports arena, on entering the stadium do you go with the flow of people, or do you try to move against it? How easy is it to walk in the opposite direction? What can you do to avoid getting trampled and being forced to change direction to go in the same direction as everyone else? When you find out a few ways to avoid a stampede, share your findings with your friends and family. It may be life-saving information.

当你去看一场足球比赛或者去任何一个体育竞技场,在进入场馆时你会随着人流走,还是试图逆行呢? 逆行是不是很不容易? 你能用什么方法来避免被踩踏,避免被迫改变方向随大流吗? 与你的朋友和家人分享可以避免踩踏的方法。这可能是救命的信息。

学科知识

Academic Knowledge

生物学	动物群的集体名称：鸟群，鱼群，昆虫群，牛群；群体智能（SI）是分散的、自我组织系统的集体行为；凯门鳄和短吻鳄之间的区别。
化　学	大脑中信息交流是化学信息从一个神经元或神经细胞发送到另一个；特定的神经元促进群体行为。
物　理	蜂拥指失控的、协调一致的奔跑，就像兽群或人群的集体冲动行为。
工程学	群体移动依据三个基本原则：单独的个人、结盟行为和凝聚的团体；在电信网络中使用群体智能；无线通信网络传输基础设施的位置；航空公司运用群体智慧，引导飞机抵达指定的机场，并引导乘客就座。
经济学	群集是计算机屏幕保护程序中的常见技术；群体效应：投资者基于一点点信息采取的行为会造成大量无根据的抛售；群体心态是金融泡沫的主要原因；经济学家可以在事后解释问题，但很难预测未来会发生什么；众筹：人们聚集在一起共同出资小额资金支持某个倡议。
伦理学	独裁者利用大众心理学强化对人民的统治，就会没有任何重大的反抗。
历　史	古罗马修辞学家昆体良在他的书《雄辩术原理》中对赢得群体思想的技巧进行了研究，对此亚里士多德在《修辞学》中也有描述；1841年，查尔斯·麦凯写了《异常流行幻象与群众疯狂》一书；1987年，克雷格·雷诺兹在计算机上用类鸟群模拟程序首次模拟了群体行为。
地　理	除了南极洲，每个大陆上都有鹈鹕存在；凯门鳄在中南美洲发现，美国和中国有短吻鳄。
数　学	"群体"的数学模型模拟了大量具有集群行为并自我推进的实体的集体运动。
生活方式	数字化工具，特别是手持式设备，促进了群体或同类相聚（乐于亲近与我们相似的人）的形成，并允许我们与同样的社会经济或种族背景下的人交往；有著名音乐家和歌手演出的露天音乐节，吸引了大量人群聚集数天在一起唱歌和跳舞。
社会学	在群体模拟中没有中央控制；人们表现出类似于群体的行为模式，如果5%的群体改变方向，其他人会跟随；人类会本能地有寻找具有相同社会经济背景或种族背景人，这强化了我们的群体偏见；从众受到缺乏责任感行为和普遍性印象的影响，这两者都随着人群的规模而增加。
心理学	行为主义者将群体行为归因于人类恐惧被孤立或惧怕失踪的自然倾向；群体心理学解释了群体的心理与个体的心理有何不同，以及它们如何相互影响；作为基本的人际关系特征，模仿可以是有意识和无意识的；群体中的人们会认为，坏事会发生在别人身上，在群体中会受到保护（数量安全）。
系统论	在群体中人的行为与个体表现不同，集体思维和行为使我们与同样背景的人一起工作，尽管加强社会凝聚的是多样性而非统一性。

情感智慧
Emotional Intelligence

凯门鳄

凯门鳄的最初看法是，逮住这些成群捕食并且都面向同一个方向的鹈鹕是很容易的事情。它观察得很仔细，注意到一些鹈鹕转向相反的方向，还有一只是把头钻进水里。凯门鳄分析指出，鹈鹕的行为像人群一样，缺乏个性。凯门鳄的生活哲学是：只有当危险发生后，才去对事件做出解释。它很好奇，向短吻鳄问了很多它不懂的问题。凯门鳄意识到，这种群体行为有利于它们，它和短吻鳄都不会被鹈鹕盯梢。 凯门鳄有个积极的愿景，希望父母承认自己的孩子会比他们做得更好。

短吻鳄

短吻鳄自信心很强，对人们的不同表现有明确的观点。当它们准备进行攻击时，它与凯门鳄分享了自己的见解。短吻鳄的想法更加消极，并且提到了人们对同胞所遭受的痛苦的漠视，以及在人群中每个人所面临的风险。它揭下虚伪面罩，承认当发生危机时，往往很少有人会挺身而出做任何事情来纠正这种情况，从而避免损害。短吻鳄促请我们根据年轻人的见解，发现他们所掌握的新知识和信息。短吻鳄的生活哲学是：寻找你的长辈没有找到的东西，并总结寻求超越每个人已经拥有的信息。

艺术
The Arts

很多人一起唱歌时，大家会有很强的归属感，例如在一个音乐节上，合唱很有感染力和魅力。但不是所有的歌曲都适合合唱，因此要选择一些适合大型团队的歌曲。尝试一下，选择一个能让更多的人参与的活动。

思维拓展
Systems: Making the Connections

群体中的人们表现和思考与个体不同。某些人有可能在不使用任何武力的情况下，就可以控制或影响人群。这可能对社会有很大的影响，特别是在宣传和传播虚假信息方面。人群控制最为突出的是在大众生产企业，它们已经使用人群控制来销售自己的产品。广告可以与宣传相比较，因为它为潜在客户准备新产品，或者激励客户重复购买同一产品，而不必进行理性的决策过程，只需要追随其他人的行为。导致吸烟和糖摄入成瘾的几种产品往往依赖于印象管理和人群控制。由于广告受到越来越多的控制，还有其他手段可以获取信息，比如通过销售具有相同品牌的户外服装，或在大型活动中免费分发饮料和糖果等。在这种情况下，重要的是，孩子们要学会独立思考，不局限于他人的想法，并且比所有同龄人想象得更多。这一点很重要，特别是在危机时期，当面临青年失业率非常高、水资源短缺和气候变化等问题时，需要找到新的方法来克服挑战。

动手能力
Capacity to Implement

你能影响一群人吗？你如何激励一个团体加入你的植树计划？你提出的论点是什么，你如何激励其他人认真对待你、尊重你的建议，并以一个志愿者的身份加入进来实施你提出的建议？当你面对反对意见时，你会说什么或做什么？通过比较其他人准备的内容，分享你对如何实现最有效运作的见解。

故事灵感来自
This Fable Is Inspired by

让·加布里埃尔·塔尔德
Jean-Gabriel Tarde

法国社会学家让·加布里埃尔·塔尔德（1843—1904），研究了集体思想（集体意识）、群体行为以及大众心理学，认为没有责任感的个人只是默默地随大流。他观察到，个体越来越多地生活在密集的城市环境中，受到许多创新事物（如灯泡、广播、摄影、电影、自行车、电话和铁路）的影响，他们形成了为新奇事物兴奋不已和追随时尚的群体。他还为经济心理学的研究奠定了基础，探究了社会和情感对个人和制度经济决策，以及市场价格和资源配置的影响。

图书在版编目（CIP）数据

冈特生态童书.第四辑：修订版：全36册：汉英对照 /
（比）冈特·鲍利著；（哥伦）凯瑟琳娜·巴赫绘；
何家振等译. —上海：上海远东出版社，2023
书名原文：Gunter's Fables
ISBN 978-7-5476-1931-5

Ⅰ.①冈… Ⅱ.①冈… ②凯… ③何… Ⅲ.①生态环
境–环境保护–儿童读物—汉、英 Ⅳ.①X171.1-49

中国国家版本馆CIP数据核字（2023）第120983号
著作权合同登记号图字09-2023-0612号

策　　划　张　蓉
责任编辑　曹　茜
封面设计　魏　来　李　廉

冈特生态童书
面向同一方向
[比]冈特·鲍利　著
[哥伦]凯瑟琳娜·巴赫　绘
高　青　译

记得要和身边的小朋友分享环保知识哦！
八喜冰淇淋祝你成为环保小使者！